W9-AUZ-145

JACKAL WOMAN

JACKAL WOMAN

Exploring the World of Jackals

LAURENCE PRINGLE

PHOTOGRAPHS BY

PATRICIA D. MOEHLMAN

CHARLES SCRIBNER'S SONS • NEW YORK

Maxwell Macmillan Canada • Toronto
Maxwell Macmillan International
New York • Oxford • Singapore • Sydney

To the next generation,
especially Matthew, Jennifer, Frank Ward, Ruth, and Amber

—P.D.M.

Charles Scribner's Sons Books for Young Readers
Macmillan Publishing Company • 866 Third Avenue, New York, NY 10022

Maxwell Macmillan Canada, Inc.
1200 Eglinton Avenue East, Suite 200, Don Mills, Ontario M3C 3N1

Macmillan Publishing Company is part of the
Maxwell Communication Group of Companies.

First Edition 10 9 8 7 6 5 4 3 2 1
Printed in Hong Kong

Library of Congress Cataloging-in-Publication Data
Pringle, Laurence P.
Jackal woman : exploring the world of jackals /
Laurence Pringle ; photographs by Patricia D. Moehlman.
p. cm.
Includes bibliographical references and index.
Summary: A photographic account of the work of Patricia Moehlman,
a wildlife biologist who specializes in the study of jackals.
ISBN 0-684-19435-X
1. Moehlman, Patricia Des Roses, date.—Juvenile literature. 2. Jackals—Juvenile literature.
3. Women zoologists—United States—Biography—Juvenile literature.
[1. Moehlman, Patricia Des Roses, date. 2. Zoologists. 3. Jackals.]
I. Moehlman, Patricia Des Roses, date, ill. II. Title
QL31.M73P75 1993 599.74'442—dc20 92-28207

Grateful acknowledgment is made to the following for the use of photographs: Dr. Sam McNaughton for the photograph of Patricia Moehlman (cover); Bjorn H. Figenschou (page 2); Dr. Marguerite Moehlman (page 5); Ira Lerner, copyright © 1972 National Geographic Society (page 7); and Dr. James Malcolm (page 26).

CONTENTS

ONE

Becoming Jackal Woman

It is midday on the Serengeti Plain of Tanzania in East Africa. The air is hot and still. A battered old Land Rover is parked in the dry grasses. Dr. Patricia Moehlman sits in it, watching, watching. Her fingers tap away on a laptop computer, recording notes about what she sees.

Vultures circle in the sky. A warthog, gazelle, or hyena may appear at any time. Patricia Moehlman studies their behavior. She has a powerful curiosity about all of nature. Mostly, however, she watches jackals.

To many people, the word *jackal* brings to mind a sneaky, low-class mammal. That is how Rudyard Kipling wrote of the jackal Tabaque in his *Jungle Book* and other Mowgli stories. Tabaque was a cunning mischief-maker and a scavenger who scrounged for food at the village dump.

Real jackals are not skulking scavengers. They do sometimes feast on the remains of animals that were killed by larger predators or that died of other natural causes. On the Serengeti Plain, an animal carcass soon attracts many meat eaters. In turn, it also at-

From her Land Rover, Patricia Moehlman studies the lives of jackals on the Serengeti Plain.

tracts tourists and other people who are interested in wildlife. They often see jackals at animal carcasses and conclude that jackals are mostly scavengers.

"Scavenging is a small part of jackal feeding behavior. Jackals are agile hunters that earn their own dinners," says Patricia.

She has observed jackals from dawn to dusk, in all sorts of weather, over a span of nearly twenty years. Patricia Moehlman is recognized as a world authority on jackals and has written about them in scientific journals, books, and such magazines as *National Geographic.* She is often asked why she chose to focus her attention on jackals.

One reason is that jackals, unlike many wild animals, do not flee at the sight or scent of a person. They are wary of humans but learn to tolerate them. "My Land Rover is like a photographer's blind," Patricia explained. "As long as I stay inside, I can approach and park within thirty yards of jackals that have grown accustomed to me and my vehicle."

An agile hunter, a silver-backed jackal leaps in pursuit of a grass rat.

Many wildlife biologists never have this experience. The men and women who study foxes, wolves, mountain lions, and many other wild animals rarely see their subjects. They often must put radio transmitters on certain individuals in order to follow their travels. Patricia relishes the opportunity to watch in solitude as wild animals go about their normal lives.

Jackals live in India and Burma, as well as in Africa, but Patricia Moehlman has been drawn to Africa since her childhood. Her father, a university professor, was fascinated with African wildlife. Patricia, her sister, and their two brothers read books and magazine articles about Africa as they grew up. "I read a lot," Patricia recalls, "and reading can affect the kind of dreams you have."

The Moehlmans lived in Washington, D.C., when Patricia was born, but soon moved to a four-acre farm near Iowa City, where her father taught at the University of Iowa. They raised sheep, pigs, and chickens. Patricia loved playing outdoors, where she sometimes pretended to be a Native American, making "Indian" camps.

In 1953, when Patricia was nine, the Moehlmans moved to a new home in the hill country west of Austin, Texas. Her mother taught high school French while her father taught history and the philosophy of education at the University of Texas. Patricia explored the wild hill country on horseback and often camped out with her sister and friends. She loved sports, especially swimming. At sixteen years of age, Patricia was co-captain of the Austin Aquatic Club swim team.

During one summer vacation, Patricia assisted in a hospital cancer-research laboratory. "I missed being outdoors a lot," she recalls. "The experience taught me how much I valued that. Right

In Iowa the young Moehlmans enjoyed farm life. Left to right: *Michael, Jacki, Stephen, and Patricia (eight years old).*

then, I suppose, I began to rule out many jobs that might have seemed interesting before."

Patricia also felt she did not want to attend a big-city college. In the fall of 1961, she began her freshman year at Wellesley College in Massachusetts. "One reason I applied there was an ecology course I saw listed in the Wellesley catalog."

However, Patricia Moehlman was fascinated by many subjects and open to a variety of careers. "My father never said, 'Girls don't do that.' I was always encouraged to figure out what I liked doing and to do it well. I was expected to pursue my own career and to support myself."

At the end of her sophomore year, she had to choose a major. Would it be art, economics, or biology? In different ways, each tempted her. But Patricia's love for the outdoors and her roots in a family that cared deeply about nature made it an easy choice: biology.

After graduating from Wellesley in 1965, she spent the summer in a training program for young biologists, studying wildlife in all sorts of habitats of the Southeast and Southwest. "We camped out all the time," Patricia recalls, "and when I returned home to Texas, I didn't want to sleep indoors."

She began work on a master's degree at the University of Texas. Her research site was Mustang Island, on the Texas coast. There she studied ground squirrels and kangaroo rats, learning about how these seed-eating rodents coexist on their dune habitat.

This project was interrupted, however, by the chance of a lifetime. Patricia learned of an opportunity to study African wildlife. Jane Goodall, famed for chimpanzee research, needed some temporary student assistants for her studies. Patricia Moehlman applied for an opening, and in 1967 she set foot in East Africa for the first time. For six months she worked with two other students and studied chimpanzee behavior at the Gombe Stream Research Center.

Jane Goodall and Hugo van Lawick wanted to write a book about three often-misunderstood African predators: hyenas, wild dogs, and golden jackals. In early 1968, Patricia helped them with jackal research in the Ngorongoro Crater. She spent several months watching a jackal family and became fond of Jason, Jewel, Cinda, and other jackals named by Hugo van Lawick. The lives of these jackals are described in the Goodall-van Lawick book *Innocent Killers*. (Published in 1971, it is still available in many libraries.)

"I can't say that I was instantly 'hooked' on jackals," Patricia recalls. "I liked them very much but was also intrigued by other

In Death Valley's Panamint Range, Patricia plays with a young orphan burro she cared for and observed.

African wildlife, including impala—the African antelopes with long, curving horns."

When the brief jackal study ended, Patricia returned to the United States to complete her studies on Mustang Island. Beyond that, she knew, lay several more years of graduate studies. One way to learn more about wild animals in their natural habitats is to pursue such studies.

In 1968, Patricia began her doctoral studies at the University of Wisconsin. Her field research, however, took place to the west, in Death Valley National Monument. She spent more than eighteen months in the southern Panamint Range of the California park, camping out and watching wild burros.

Also called donkeys or asses, these burros are descendents of animals brought to western North America centuries ago by the Spanish. Burros that escaped or were turned loose formed wild bands. In a one-hundred-square mile area of Death Valley, Patricia observed more than two hundred burros in order to learn more about their social lives and how they survived in this arid habitat. She concentrated her work in a high-desert area called Wildrose Canyon.

Patricia traveled by jeep, horse, and on foot, often walking ten to twenty miles a day. Many days her observations began at 5:30 in the morning as she used binoculars to study burro behavior, noting their social interactions and what plants they ate. After dark, she sometimes watched burros with a special night-vision scope. Every day she wrote detailed notes of her observations. They were the basis of her doctoral thesis, which was supervised by Dr. John Emlen at the University of Wisconsin.

During her study of burros, Patricia had the companionship of a Siberian husky, a female named Tasha. "She was a friend and also a tutor. Tasha introduced me to dog behavior. Watching her, I became more sensitive to the subtle facial expressions and body postures of canines. I didn't know it at the time, but Tasha helped prepare me for my observations of jackals."

When Patricia's fieldwork ended, in 1972, she missed the individual burros she had come to know and the sound of the males braying in the desert morning. "I also missed the solitude and the satisfaction of camping in open country, where I could see for miles in all directions."

In 1974 she again had that solitude and satisfaction—in East Africa. Camped on top of a hill, she looked over Lake Lagarja to the north and many miles in all directions. With funds from the National Geographic Society and the Harry Frank Guggenheim Foundation, Patricia began to watch jackals.

TWO

Getting to Know Jackals

The vast game preserve of the Serengeti Plain is an ideal place to study jackals. Patricia Moehlman's camp is located where two species live—golden jackals in the short grasslands and silver-backed (sometimes called black-backed) jackals in the nearby bush woodlands.

When she is asked to describe jackals, Patricia's voice rises with enthusiasm: "They're more beautiful than most people expect. A golden jackal looks very much like a small coyote. Picture an animal that weighs about fifteen to twenty pounds, with an appealing, wolfish face and a golden coat—sometimes more brown, sometimes more gray—but basically a golden tawny.

"Silver-backed jackals have reddish fur except for silvery hair on their backs, a foxy face, large ears, and very long legs. They're great leapers.

"Once people really watch jackals, they can appreciate their beauty. And once you know how they live, their ways of living are admirable."

Patricia began to study families of silver-backed jackals that

Left, *a golden jackal father with his pups;* right, *a mother silver-backed jackal nursing hers.*

lived in the bush woodlands surrounding Lake Lagarja. Their territories were a short drive from her camp. In fact, the campsite she chose was within the boundaries of one jackal territory.

Over a span of several years, Patricia kept track of more than a dozen pairs of silver-backed jackals. She also began to observe pairs of golden jackals in the grasslands about nine miles from her camp. In describing jackal behavior, the word *pair* has special significance. Jackals, which live about six to eight years, usually pair for life. They are among a small fraction of all mammals that do so.

The strong bond between a pair of jackals is plain to see. Jackals nuzzle their mates affectionately. They groom one another, nibbling fur, for several minutes at a time. The strong bond between male and female is further demonstrated when they raise pups.

Silver-backed jackal pups are usually born in July and August—during the dry season of the Serengeti. Golden jackal pups are born in December and January, when the rains begin. For several weeks the pups are out of sight in the family den, nursed and groomed by their mother. The father brings food to her.

Jackals don't usually carry food very far in their mouths. In-

A female silver-backed jackal grooms her mate.

stead, they swallow it. Once they are safely home, they cough it up. This regurgitation is how male jackals usually feed their mates.

"Regurgitated food does sound a bit unpleasant," says Patricia Moehlman, "but in fact it is a very efficient way for the adults to transport food. The jackals must not only catch prey, but they must retain it and keep other predators from stealing it."

Once the pups start to eat solid food, they, too, get most of it by regurgitation. The pups first wobble out of their den when they are about three weeks old. As the days pass they grow steadier on their feet and spend more time aboveground, playing.

Golden jackal pups greet their father and beg him to regurgitate food he has carried back to the den.

Golden jackal pups at play.

"At first," Patricia Moehlman says, "the games are clumsy attempts at wrestling and pawing and biting. As they become more coordinated, they ambush and pounce and chase each other. They play tug-of-war with a bone or an ostrich feather." Parents are very playful, too. Games of chase are popular. Especially when only one pup survives, the adults become enthusiastic play partners.

Patricia has found that hyenas, leopards, eagles, lions, and other predators try to kill young jackals. "Many times," she recalls with glee, "I have seen a fifteen-pound jackal chase a one-hundred-twenty-pound hyena away from its den. The quick and agile jackal will run up behind the more clumsy hyena and bite it in the rear, darting away before the hyena can even turn around. Often all a jackal has to do is bark, and a hyena will retreat with its tail between its legs."

A golden jackal father bites and chases a spotted hyena away from his family's den.

In an attempt to keep her pups safe, a mother silver-backed jackal changes dens about every two weeks. Silver-backs may have as many as eight dens in their territory. "She whimpers to her pups as if about to feed them and then leads them to a new den. Sometimes a pup gets left behind and the father stays with it until the mother returns."

Pups that survive for about fourteen weeks are strong enough to have a good chance of becoming full-grown adults. As the pups grow, they begin to follow their parents when the adults forage throughout their territory. They watch their parents hunt and try it themselves. By the time the pups are six months old, they are hunting on their own successfully.

Silver-backed jackals catch many grass rats—small rodents that are plentiful in the dry season. They also eat mice, fruits, hares, insects, reptiles, and small gazelles. Food is abundant all year round in their brushland habitat, but the most plentiful items—rats, fruit—are small, so silver-backs often travel as far as five miles a day in their hunts. Each family has a territory of two to three square miles.

As daylight fades, the family reassembles. The bushy land-

Young jackal pups first emerge from the den at about three weeks of age.

A silver-backed pup holds a dead grass rat that he has just received from his mother.

scape prevents silver-backs from seeing far, but their yipping calls help the family members reunite. Each jackal recognizes the calls of its own family and ignores those of jackal neighbors.

Golden jackals have smaller territories. Their time of greatest food need—when they have pups—is the rainy season. The lush grasslands are alive with prey.

Patricia Moehlman describes the change of seasons: "In the dry season, from June through October, your footsteps or vehicle wheels stir up a fine dust from the volcanic-ash soil. The grass is dry and yellow. As far as you can see, only a few gazelles, and maybe a warthog and some oryx, are visible. Golden jackals survive the dry season by digging up lizards and dung-beetle balls or catching an occasional snake.

"When the rains come, within three days' time the grass springs up from its roots and the land becomes lush green. Thousands upon thousands of wildebeests, Thompson's gazelles, and zebras appear. It is their time to feast on grass and to give birth to their young."

The jackals themselves kill young gazelles and eat the afterbirths (placentas) of wildebeests and other large grazing animals. This is also their time to be scavengers of carcasses.

Golden jackals scan the sky for circling vultures, then trot to the animal remains that attract these birds. They may only have a few moments to feed before a hyena, lion, or a dozen large vultures claim the prize. This is a risky situation, as a jackal tries to get some food-to-go while in close range of vulture beaks or hyena jaws.

The teamwork of jackals, and their speed and agility, often help them get more food. One jackal nips at a hyena's rear or ankle. The hyena wheels around and chases the jackal briefly, giving a second jackal an opportunity to dash to the carcass.

A golden jackal defends a carcass against many vultures.

Jackals also cooperate in hunting live prey. One jackal draws the attention of a Thompson's gazelle away from her young. Its mate rushes in to attack the young gazelle.

The equal relationship between male and female jackals is rare among mammals. It is shown even in the pair's defense of their territory. When a strange jackal is sighted near the territorial border, usually both male and female trot toward the intruder. Through scent or some visual clue (only the jackals know), they can tell at a distance whether the stranger is a male or female. If it is male, the resident male challenges it; if the intruder is female, the resident female defends the territory.

In many mammal species, fathers help little or not at all in raising the young. In contrast, jackal fathers help enormously in rearing their offspring. Jackals are medium-sized predators, with

A male silver-backed jackal watches as his mate threatens a female who intruded into their territory.

Without the help of their fathers, few jackal pups would survive to become adults.

many larger enemies, trying to survive in a sometimes-harsh environment. In the Serengeti, the survival of pups often depends not only on both parents but on jackals that Patricia calls "helpers."

A helper is a young jackal that stays with its parents for an extra six to eight months. When a jackal is ten or eleven months old, it is a young adult, ready to leave home, seek a mate, and establish a territory. Some young jackals do this, but others stay. They assist their parents in rearing the next litter of pups. From her observations, Patricia Moehlman has concluded that helpers often play a vital role in the survival of their little sisters and brothers.

One example occurred early in her studies. In 1976–1977 there was a high population of grass rats in the bush woodland around Lake Lagarja. Food was plentiful for silver-backed jackals. A pair Patricia named Tipper and Tamu had a litter of five pups in June 1977. Tipper and Tamu were experienced parents. Food was abundant. Nevertheless, none of the pups survived. Tipper and Tamu had no helper.

At the same time, another jackal pair, Scorpio and Libra, lived in a nearby, similar-sized territory. They had a helper, Orian, from their previous litter. Even though grass rats were less plentiful than in the territory of Tipper and Tamu, these jackals succeeded in raising three of their six pups. The teamwork of three adults, including Orian, seemed to make the difference. Patricia concluded that "Food resources are a critical factor in the survival of pups, but more important is the number of adults in the family that can capture the food and supply it to the pups."

In some ways, helpers are treated like—and act like—pups. They do not mate or mark territory borders with urine. They do not challenge the authority of their parents.

In other ways, however, they act like the capable young adults they are. "These older brothers and sisters," says Patricia Moehlman, "are not just hanging around the territory. Helpers hunt and bring food home to their mother while she is nursing the new pups. They also regurgitate food to the pups, play with them, teach them to hunt, guard them, and groom them.

"More pups survive if there is an adult guarding the den.

With a helper standing guard, the parents can spend more time hunting. This allows the most experienced hunters to spend more time foraging."

Parent jackals sometimes share food with helpers. They groom and play with them. The strong social bonds of the jackal family are maintained. Each jackal seems to care about the well-being of the other family members.

Patricia Moehlman has seen countless examples of these strong family bonds, but one example stands out. In February 1977, a pair of golden jackals, Raha and Refu, had three pups about nine weeks old. The parents and their helper, Ngumu, had kept them well-fed and the pups were doing well. Then came four days of heavy rain and cold weather. When the sun finally shone, only one pup, Safi, was barely alive.

"She lay shivering near the den entrance. Raha groomed her and gently touched her with a paw. Safi, barely lifting her head, snarled. Refu tried to play with Safi but got no response.

"All three adults left to hunt and killed a young Thompson's gazelle. Refu was the first to finish eating. He hurried back to feed Safi. When he arrived, she was dead. He nuzzled her and whimpered. Finally he picked her up in his mouth and carried her a quarter of a mile away, where he dug a hole and buried her. He pushed loose soil with his nose to cover her.

"Refu then trotted back to the den. Raha and Ngumu had arrived there and were searching for Safi. When Refu was a few yards from them he stopped and howled. So did Raha and Ngumu, in a doleful way I had never heard before."

Patricia Moehlman says, "I can only surmise that they were mourning their dead pup."

A year-old silver-backed helper looks after his younger brother.

THREE

Watching Closely, Asking Questions

Around Christmastime, snow may be falling in Connecticut, where Patricia Moehlman has a home. She, however, is usually seven thousand miles away, in the snowless Serengeti. She wouldn't want to miss another pup-rearing season of individual jackals she has known for years.

Patricia divides her time between Tanzania, the United States, and sometimes other places. In 1980, for example, she was asked to spend several months studying a population of wild burros on one of the Galápagos Islands, about six hundred miles west of Ecuador. That same year and in part of 1981, she was a visiting scientist at Cambridge University in England. Also, for five years she taught courses in wildlife ecology at Yale University in Connecticut. Since 1986 she has been a research scientist for NYZS—The Wildlife Conservation Society.

As often as possible, she returns to her campsite on the hill overlooking Lake Lagarja. A large tent, with bed, table, chairs, and cooking utensils, is left standing. Patricia buys food locally

Patricia's camp sits among acacia trees on a hill overlooking Lake Lagarja.

and thrives on the fresh fruit and vegetables that are available.

"In the Serengeti," she says, "I lead a very simple, very basic life but have learned a few tricks to give myself treats. For example, I can bake bread or even a chocolate cake in a campfire, under an upside-down metal bowl covered with hot coals."

There are hardships. For several years, gasoline supplies were limited in Tanzania. In order to save fuel, Patricia sometimes slept in her car near the golden jackals, rather than drive back to camp. Nearby Lake Lagarja is a soda lake, with a high concentration of sodium carbonate that makes its water unfit for drinking. Patricia

Jackal research in Tanzania has taught Patricia many skills, from baking cakes in campfires to working on Land Rovers.

has to fetch fresh water with her trusty old Land Rover. And to ensure that her vehicle *is* trustworthy, she has learned to be an auto mechanic.

Patricia is often asked whether she feels lonely. She says, "Certainly there are times I miss my family and friends, but I need a fair amount of solitude. There's a big difference between solitude and loneliness. A person can feel lonely in a crowded city. I spend a lot of time alone, but it is not an enforced isolation. It is always my choice. I can hop in the Land Rover and visit friends, or invite them to my camp. There are people around, including local nomadic cattle herders (Maasai) and other scientists doing research in the area.

"I believe that people need to feel that they are part of a community. It is not a matter of numbers but of support and co-operation, which I have always had in Tanzania."

She rises before dawn, has a quick breakfast, then drives to the rendezvous area of one of the jackal families she is studying. Sometimes she takes photographs or makes drawings of the jackals. These help her identify individuals, a key part of her research.

Scientists sometimes have to capture individual animals and mark them in some way (numbered or colored tags, paint marks) in order to recognize them later. When Patricia Moehlman began studying jackals, she hoped to avoid that. "I wanted to observe them in their natural habitat with as little human interference as possible. Looking closely at them, I began to notice natural markings that helped me tell one jackal from another.

"The male I called Refu, for example, had a scar just under his right eye. His mate, Raha, had a slit in her right ear. Another jackal I named Tipper because he had a white tail tip."

Patricia often gives the jackals Swahili names. She has watched some pairs survive, through good times and bad, for as long as seven years. She grows fond of them. She cares about what happens to them. At times, she is tempted to interfere: "It is hard to just sit and watch when pups are sick."

But she keeps a distance and has never tried to tame a jackal. "That might be fun for me," she says, "but not good for them. First of all, it would be bad science to intrude in their normal lives that I am trying to understand. Second, I feel a responsibility

Raha had a slit in her right ear, a mark that helped Patricia recognize this golden jackal.

to them that goes beyond science. I respect and appreciate jackals for what they are."

Although Patricia watches jackals from dawn to sunset, and sometimes through the night for twenty-four hours, she is humble about her ability to understand their lives fully: "I am limited by what kind of animal I am. I can't hear everything jackals hear. I can't see everything jackals see. And I certainly can't smell everything jackals smell. But with time, sitting still in the Land Rover or following them in it, I keep learning more.

"I learn something new every day about the jackals and their environment. The focus is on jackals, but everything that goes on is part of the study too."

When food supplies are adequate, golden jackal parents are exceptionally gentle and loving toward their pups.

Patricia is a behavioral ecologist. One aim of her studies is to see how an animal's behavior is affected by its environment. Scientists have learned that many animal species do not have just one, unchangeable social system. Their social behavior varies with the environment in which they live.

Patricia witnessed a striking example of this among golden jackals. After several years of studying both goldens and silverbacks, she wrote in the December 1980 issue of *National Geographic*: "Golden jackals tend to be friendlier with their pups than do silver-backed jackals. Their greetings are gentler . . . golden parents are more tolerant of their young and allow pups to crawl all over them and play tug-of-war with their tails."

She came to these conclusions after watching golden jackals raise their pups in wet seasons, when food was plentiful. Then, in 1982, she saw for the first time a family of golden jackals trying to raise pups in the dry season. Food was scarce. She saw the parents and helpers exhibit threats and aggression toward the demanding pups. The environment had changed, and with the change in food resources much of the affectionate family life that she had observed before vanished.

Even with plentiful food, the wet season can be a tragic time for golden jackal families. Most of the families include one or more helpers that guard pups and bring them food. Nevertheless, few pups survive. The parents and helpers do their best, but their efforts are often overwhelmed by circumstances beyond their control. When heavy rains flood their dens, many pups die of exposure and the resulting illness.

The presence of helpers in jackal families is unusual among mammals. The lives of helpers, and their role in their families, intrigues Patricia. Among silver-backed jackals, the family clearly gains with the aid of a helper: more pups survive. And yet, only

Golden jackal pups try to dry out after a rainstorm. Many pups die when they become wet and cold in the rainy season.

about one out of four surviving pups stays to become helpers. Why don't more stay? The answer, Patricia Moehlman believes, lies in the environment of silver-backs. Food is abundant all year round. Also, it may be possible in the bush woodland habitat for a young adult jackal to avoid detection and set up a territory. Thus, a young silver-back need not delay its urge to mate and begin raising its own young.

The environment of golden jackals is different, and so is the behavior of the young adults. Golden jackal pups that survive leave their home territory during the dry season, when food is in

When silver-backed pups grow up, they usually strike out on their own and have a good chance of finding mates and setting up new territories.

short supply. Many of them return in the wet season, rejoin their parents, and become helpers.

Humans might call this kindhearted, altruistic behavior. In the world of jackals, however, this behavior is simply a response to the environment. The open Serengeti Plain is a mosaic of golden jackal territories. It is very difficult for a young adult golden jackal to find a mate and acquire a territory. So most return home. As helpers, they gain experience hunting and caring for pups. When they do leave home for good, helpers may have a better chance of mating and raising their own pups than those that do not stay to help.

Patricia Moehlman has seen evidence that some young jackals may not have the choice to stay or leave. In a litter of pups, some

A golden jackal with a dung ball, which contains beetle larvae—a valuable source of food in the Serengeti dry season.

Patricia will use radio-tracking devices to follow golden jackals, like this juvenile, to learn how they find a mate and get a territory.

are dominant, others more submissive. If a dominant pup chooses to stay with its parents as a helper, it may force one or more of its more subordinate brothers or sisters to leave.

Patricia has many questions about both the helpers and the individuals who leave. Unfortunately, she can't answer all of these questions by quietly watching. The only way to keep track of individuals that leave is to attach radio transmitters to them, then to locate them by using a special radio receiver.

This technique has been used to study many wildlife species.

However, Patricia has been reluctant to use it. "For me, that's a dilemma. Radio collars can affect survival. There's evidence that animals carrying them have a higher mortality than normal. The extra weight may affect their running or fighting ability. In a bush or woodland environment, there's a chance of the collar getting caught on something. Also, I would be interfering in jackal lives, because radio tracking involves catching and handling jackals, putting radio collars around their necks, then releasing them."

Aiming to do as little harm as possible, she has studied radio tracking carefully. She explored the possibility of a radio transmitter implant, rather than a collar. That was ruled out because signals from such transmitters do not travel far enough for her to keep track of jackal movements. Also, she is reluctant to use the dart guns that are commonly used to inject drugs (which temporarily knock out animals). The darts may do harm, even breaking a bone, in an animal the size of a jackal.

Patricia will investigate the use of a blow dart or the effectiveness of putting the drug in bait. She has also sought a radio collar that would be best for her study and for the jackals. Ideally, its batteries would last for at least six months, and the collar would be fastened so that it drops off after ten months. It needs to be small and light so that it will be covered by the ruff of fur around the jackal's neck.

Patricia Moehlman is excited about this next step in her research. Nevertheless, her hunger to learn about the lives of jackals is tempered by her concern about those same lives.

FOUR

Conserving Serengeti's Wildlife

Each year as the rainy season starts, Patricia sees the arrival of the migration. Vast herds of wildebeest, zebra, and gazelle arrive in the shortgrass plains to give birth and raise their young. "It is impossible," she says, "to express in words the fundamental majesty of hundreds of thousands of animals spread to the horizon."

Events like this remind Patricia Moehlman of how lucky she is, to be outdoors, observing wild animals of the Serengeti Plain—one of the most extraordinary wildlife habitats on earth. As much as she enjoys her research, however, she knows that some of her time and effort must be aimed at helping to save all life of the Serengeti.

During her years there, she has been impressed with the Tanzanian people's commitment to conserving their natural resources, their *mali hai* or "living wealth." The United Republic of Tanzania has designated over 25 percent (90,750 square miles) of its land for conservation areas. This is a huge undertaking, both scientifically and economically.

With Tanzanian wildlife managers, scientists, and local inhab-

The rainy season brings vast herds of grazing animals to the Serengeti Plain.

itants, Patricia is trying to find ways in which both people and wildlife can continue to coexist. She also teaches and trains Tanzanian citizens to conduct ecological studies. As a research scientist for NYZS—The Wildlife Conservation Society, she helped set up ecological monitoring programs in the Ngorongoro Conservation Area and in three Tanzanian national parks.

Long-term ecological research is needed to understand and manage these areas. Rainfall is a critical component of these ecosystems and often determines how much grass can grow and how long wildebeests and other grass eaters can use an area. Predators

From mighty lions to fifteen-pound jackals, many predators depend on the Serengeti's herds of grazing animals.

ranging in size from jackals to lions depend on the grazing animals for their survival and reproduction. The monitoring studies that have been set up measure rainfall and other weather data, plant growth and its use by animals, and the movements and numbers of plant eaters and their predators.

Patricia says, "These long-term studies will help us to track and understand changes in these ecosystems—changes in climate and in plant and animal populations and their relationships. The information will also help us to monitor changes caused by humans."

In 1988–1989, Patricia's concern about wildlife conservation took her to Somalia, where she surveyed a dwindling population of the African wild ass. Camel herdsmen in Somalia and neighboring Ethiopia often shoot them on sight. This relative of the

"Jackals have lives that humans might want to emulate."

wild burro of the American Southwest is the earth's most endangered member of the horse family. Patricia is involved in efforts to save the African wild ass from extinction; she would also like to study its behavior.

"I may study other animals, or jackals in other habitats. But my work will be basically the same—observing wild animals in natural habitats, trying to understand how their behavior is influenced by the ecological setting."

Books, magazine articles, films about jackals on public television—all are helping to change notions about jackals. The "skulking scavenger" image erodes as people learn about the lives of real jackals.

Humans appreciate the kind of behavior that is the core of jackal social life. Patricia says, "If I believed in reincarnation—which I don't—I would be happy to come back in a future life as a jackal. They lead such generous lives.

"Jackals have lives that humans might want to emulate. They have enduring and loyal pair bonds; they share food and care for younger, sick, or injured relatives; they are courageous in their defense of home and family."

Through the efforts of Patricia Moehlman and other wildlife biologists, people are finally discovering that jackals are creatures to be respected and admired.

FURTHER READING

Moehlman, Patricia D. "Ecology of Cooperation in Canidae." In *Ecological Aspects of Social Evolution*, edited by D. Rubenstein and R. Wrangham. Princeton, N.J.: Princeton University Press, 1986.

———. "Getting to Know the Wild Burros of Death Valley." *National Geographic,* April 1972: 502–517.

———. "Jackals" and "Jackal Helpers." In *The Encyclopedia of Mammals,* edited by D. W. MacDonald, Oxford, England: Equinox, 1984.

———. "Jackals of the Serengeti." *National Geographic,* December 1980: 840–850.

———. "Social Organization in Jackals." *American Scientist,* July–August 1987: 366–375.

Van Lawick-Goodall, Hugo and Jane. *Innocent Killers.* Boston: Houghton Mifflin Co., 1971.

INDEX